Earth's Oceans

SCHOOL PUBLISHERS

Orlando Austin New York San Diego Toronto London

Visit *The Learning Site!*
www.harcourtschool.com

Lesson 1

What Are the Oceans Like?

VOCABULARY

salinity
water pressure
continental shelf
continental slope
abyssal plain

The amount of salt in water is called **salinity**. The Dead Sea, shown here, has very high salinity.

Water pressure is the weight of water. Water pressure increases as you go deeper underwater. The small cup was crushed by water pressure.

The part of the ocean floor that drops gently from the land is the **continental shelf**. The part of the ocean floor that drops steeply is the **continental slope**. **Abyssal plains** are the large, flat areas of the ocean floor.

READING FOCUS SKILL

MAIN IDEA AND DETAILS

The **main idea** is what the text is mostly about. Details are pieces of information about the **main idea**.

Look for information about the conditions and features of the ocean and then note **details** about each condition or feature.

Ocean Water

Ocean water is salty. The salt in ocean water comes from minerals on land. The amount of salt in water is called **salinity**. The salinity of the ocean does not change much from place to place. A lake in Israel, called the Dead Sea, is tens times as salty as ocean water. The Dead Sea has high salinity because more water evaporates from it than flows into it. As the water evaporates, salt is left behind.

▼ **The Dead Sea**

The submersible *Alvin* has thick walls that protect it from water pressure.

Temperature in the ocean varies from place to place. Temperature is higher at the surface because the sun warms the water. But most of the ocean water is deep. About 90 percent of the ocean has very cold temperatures.

The deep ocean is also very dark. The water absorbs and scatters light. As you move deeper, it becomes darker. At some point you can see no light at all.

Water pressure is the weight of water. The deeper you go, the greater the water pressure. This is because more and more water pushes down. The pressure becomes so great that it can collapse submarine walls. Scientists use submersibles to study the deep ocean. Submersibles are small submarines with thick walls. The thick walls keep the water pressure from collapsing the sides of the submersible.

What condition does not change much from the suface of the ocean to the deep ocean?

The Ocean Floor

The ocean floor has many features that are like land features. It has mountains, valleys, and volcanoes.

The ocean floor drops gently away from the land. This part of the ocean floor is the **continental shelf**. Then the ocean floor drops steeply. This part is the **continental slope**. The continental slope has deep canyons cut into it. They were formed by rivers flowing into the ocean. The continental rise is at the base of the continental slope. Sediment from rivers moved into the ocean and piled up there to form the continental rise.

Large, flat areas of the ocean floor are called the **abyssal plains**. In some areas, deep trenches cut into these plains. In other places, mountains rise up. Islands in oceans are actually the tops of underwater mountains.

Where is the continental rise and what materials form the continental rise?

▼ **Features of the Ocean Floor**

Changes to the Ocean

The ocean is always changing. Islands in the ocean form and then disappear. Off the coast of Iceland, a volcano erupted underwater in 1963. It began building up layers of lava. After six months the volcano reached the surface. It formed an island, Surtsey. Once the volcano stopped erupting, Surtsey began to get smaller. The island has sunk a little and the waves are wearing away its rock.

The island of Bora Bora is also disappearing. The island is slowly sinking. The ocean is wearing away the rock. Someday, the only reminder of Bora Bora will be the coral reef that surrounds the island.

How can islands form in the ocean?

▲ **The Island of Surtsey formed from underwater volcanic eruptions.**

Review

Focus Skill

Complete this main idea statement.

1. ______ is the downward push of water that increases as depth increases.

Complete these detail statements.

2. The ______ is the gentle, sloping part of the ocean floor.
3. Mountains and trenches are found on the ocean's ______ ______.
4. ______ ______ form islands in the always changing ocean.

Lesson 2

How Does Ocean Water Move?

VOCABULARY
wave
current
tide

A **wave** is the up-and-down movement of water. Waves carry energy across the ocean.

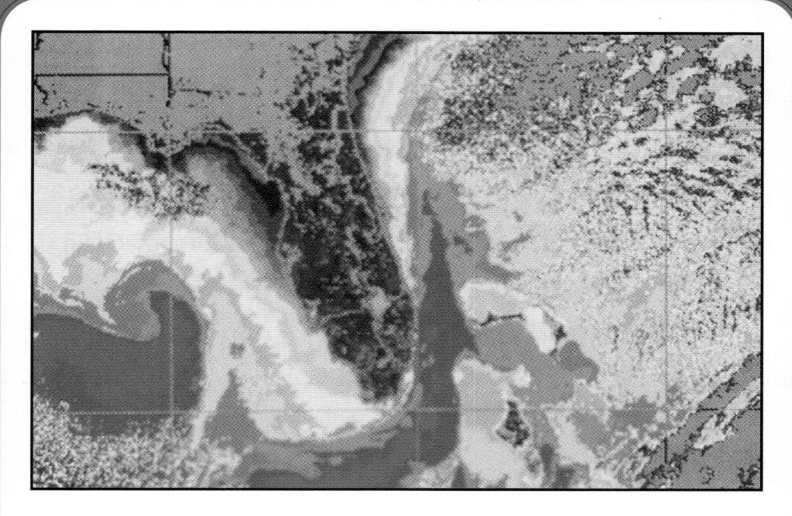

A **current** is a stream of water that flows through the ocean. Large currents, like the ones shown in the picture, carry water great distances.

A **tide** is the rise and fall in the water level of the ocean.

READING FOCUS SKILL
COMPARE AND CONTRAST

You **compare and contrast** when you look for ways ideas in the text are similar and different. **Compare** means to find the way things are similar. **Contrast** is to find the way things are different.

Compare and contrast how ocean water moves.

Waves

Have you ever seen ocean waves? It looks like water is moving across the ocean in ridges. But it is not. A **wave** is the up-and-down movement of water.

Air blowing across water makes waves. Energy from the moving air moves the water. A wave carries energy across the ocean. But the water in a wave moves in a circle. As a wave moves forward, only the energy moves forward. The water stays in the same place.

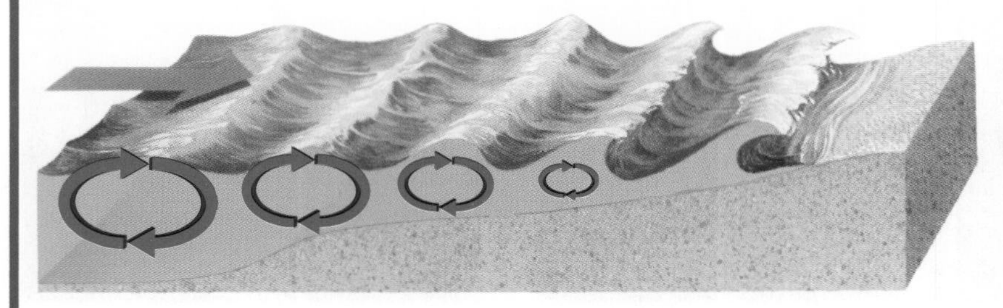

As a wave approaches the shore, the bottom of the wave slows down and the top falls foward.

▲ **Breaking Waves**

When water becomes shallow, waves slow down. The bottom of the wave slows more than the top. This causes the wave to fall over, or break. Breaking waves are what you see when the water crashes onto the beach.

A storm surge is a mound of water pushed to shore by a hurricane. A storm surge can cause a lot of destruction.

An earthquake or volcanic eruption can make a *tsunami*. A tsunami is a long, fast moving wave that can cause a lot of destruction when it reaches land.

Compare and contrast a storm surge and a tsunami.

▼ **Tsunami damage**

Currents

Did you know that the sun causes the ocean to move? The sun's energy heats air around the equator. This air then moves toward the poles. As it moves, the air pushes the water. This causes a **current**. A current is a stream of water that flows like a river through the ocean.

Surface currents are large currents. The Gulf Stream is a surface current that moves across the Atlantic Ocean.

The ocean also has small currents. A longshore current moves along the shore. Swimmers may get into the water in one spot along the shore. Later they find themselves farther down the shore. They have been moved by a longshore current.

A rip current carries water away from the beach. A rip current is dangerous for swimmers, because it moves very quickly away from shore.

▼ **The gap in the water shows the rip current.**

◀ **The satellite photo shows the large area in which an El Niño formed.**

Currents can also flow deep in the ocean. In the ocean near South America, winds blow warm surface water away from the land. Deep ocean currents then carry cooler water up to the surface.

Winds can affect these currents. If the winds change direction, the warm surface water stays where it is. Then deep, cold currents do not move to the surface. This causes a change in the weather called *El Niño.*

El Niño changes the weather because warm water evaporates faster than cool water. This causes more clouds and more rain. The wind also pushes the warm water to the west. The *El Niño* weather pattern brings more rain to coastal areas.

Contrast a longshore tide and a rip current.

The rains from El Niño ▶ can cause a lot of damage.

Tides

The level of the ocean rises and falls each day. This rise and fall of the water level is a **tide**.

Tides are caused by the force of gravity from the sun and moon. They pull on Earth's oceans. The pull from the moon affects tides more because the moon is closer to Earth.

The pull of the moon on Earth causes two bulges of water to form. One bulge is on the side of Earth facing the moon. The other bulge is on the opposite side of Earth. The level of the ocean is highest at these bulges. This causes a *high tide.*

Tides are higher when the sun and moon line up. This is a spring tide. ▼

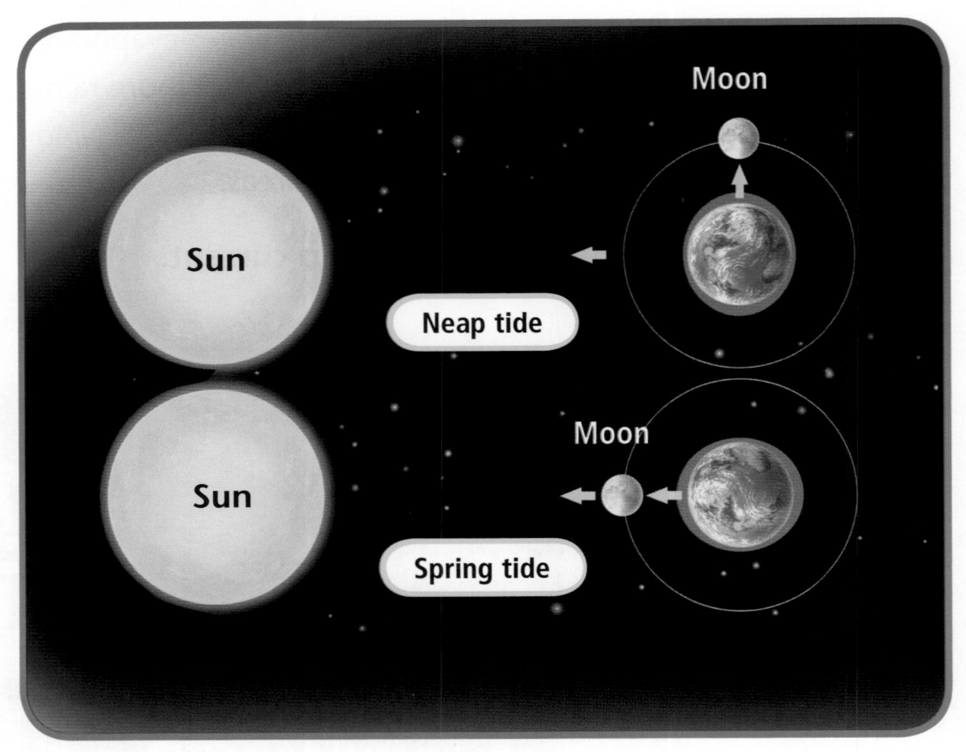

A *low tide* occurs between the two bulges. This is where the level of the ocean is lowest. Each day a beach will usually have two high tides and two low tides.

Compare and contrast high tides and low tides.

A beach at high tide. ▶

Review

Focus Skill

Complete these compare and contrast statements.

1. A wave pushes ______ forward while water stays in almost the same place.
2. A ______ current moves away from the shore while a longshore current moves along the shore.
3. A ______ tide occurs when a beach is either closest to or farthest from the moon.
4. ______ can be caused by earthquakes while storm surges are caused by hurricanes.

Lesson 3

What Forces Shape Shorelines?

VOCABULARY

shore
headland
tide pool
jetty

The area where the ocean and the land meet is called the **shore**. A **headland** is a point of land left behind after soft rock wears away.

A **tide pool** is a temporary pool of ocean water. A tide pool is formed when water is trapped after the tide goes out.

A **jetty** is a rocky structure that sticks out into the ocean. This structure blocks currents from carrying away sand.

READING FOCUS SKILL

MAIN IDEA AND DETAILS

The **main idea** is what the text is mostly about. Details are pieces of information about the **main idea**.

Look for information about what forces shape shorelines and **details** about each force.

Shore

Have you ever been to the shore? Some shores are covered with sand. Other shores are rocky. The **shore** is the area where the ocean and the land meet.

At shores covered with sand, the water can carry the sand away. At rocky shores, the water can wear down the rock. Some shores had both hard rock and soft rock. After the soft rock has worn away, hard rock is left behind. A **headland** is a point of land formed this way.

A sandy shore and two headlands

Waves slowly wear away rock that forms a headland. Over time, the waves can form caves. Sometimes the waves wear all the way through the rock. This forms a hole called a *sea arch.*

Some shores have areas that are underwater at high tide, but exposed at low tide. **Tide pools** form when small pools of water are trapped at low tide. Animals can get trapped in tide pools.

Estuaries are places where a river flows into an ocean. These areas have many kinds of plant and animal life. Estuaries are also homes to many kinds of fish.

What causes tide pools to form?

Human Activities Affect the Shore

Sand is washed away from beaches every day by waves and currents. This sand may be deposited at other places. Longshore currents often move sand. This movement causes difficulties for people. It makes some beaches smaller while others grow larger. The sand also fills in channels that boats move through.

One way people fix this problem is by removing sand from a channel and putting it on a beach. They do this with pumps, bulldozers, and other large equipment. This is called *beach restoration.* Beach restoration is very expensive. It can cost millions of dollars. It is also not a permanent fix to the problem. More sand must be added to the beach every few years.

The bulldozer is working to restore the beach.

Because replacing sand on a beach is expensive, people have developed ways to stop the loss of sand from beaches.

▲ **This jetty helps prevent currents from eroding the sand at the beach.**

One way is to build **jetties**. These are wall-like structures that stick out into the ocean. They are usually made of large rocks. Jetties block currents and keep sand on the beaches. But a problem with jetties is that they keep sand from being carried to other beaches.

Coral reefs near coasts naturally absorb a wave's energy before it reaches the shore. This helps protect the shore. People build artificial reefs with many different objects. Most reefs are built to provide homes for fish and other sea life. Protecting the beaches is an added benefit of artificial reefs.

What is the main reason why people build artificial reefs?

◀ **Artificial reefs are made of many different objects.**

Mysteries of the Oceans

The oceans are so deep and wide that people cannot easily explore them. But for many years people have wanted to learn more about the oceans.

Scientists have developed many ways to explore the ocean. Diving suits, diving bells, and submarines have helped scientists study the deep sea. Scientists have even developed remote-controlled vehicles. These have attached cameras that let them explore areas too deep for humans to explore.

Diving suits and diving bells have helped people explore the deep ocean.

Explorers have discovered many interesting things during deep sea exploration. *Hydrothermal vents* are cracks in the ocean floor. They release hot water and minerals into the ocean. Large worms have been found living near these vents. These worms are up to 3 meters (10 feet) long!

Scientists continue to explore the oceans. More exploration will help scientists understand many of the ocean's mysteries.

What keeps people from simply swimming deep underwater to study the ocean?

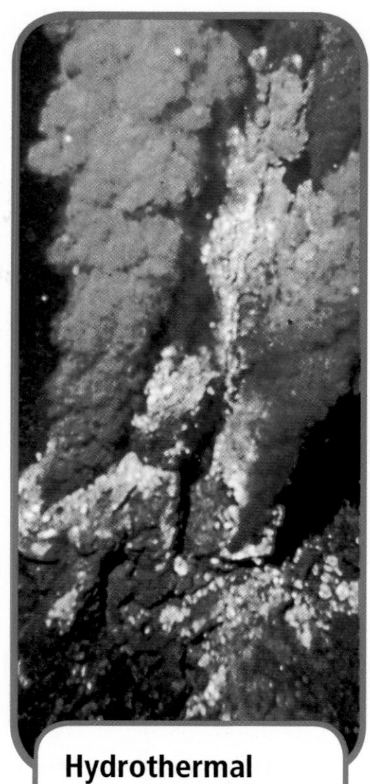

Hydrothermal vents are cracks in the ocean floor.

Review

Complete this main idea statement.

1. ______ erode sand from beaches and deposit it along the shore, in channels, or on the ocean floor.

Complete these detail statements.

2. A ______ keeps one beach from losing its sand but prevents sand from being deposited at other beaches.
3. A ______ is formed when soft rock is worn away and hard rock forms a point of land.
4. Worms have been found living near ______ ______ in the ocean floor.

GLOSSARY

abyssal plain (uh•BIS•uhl PLAYN) a large, flat area of the ocean floor.

continental shelf (kahnt•uhn•ENT•uhl SHELF) the part of the ocean floor that drops gently near the land.

continental slope (kahnt•uhn•ENT•uhl SLOHP) the part of the ocean floor that slopes steeply.

current (KUR•uhnt) a stream of water that flows like a river through the ocean.

headland (HED•luhnd) a point of land at the shore where hard rock is left behind and other materials are washed away.

jetty (JET•ee) a wall-like structure that sticks out into the ocean to prevent sand from being carried away.

salinity (suh•LIN•uh•tee) the amount of salt in water.

shore (SHAWR) the area where the ocean and the land meet and interact.

tide (TYD) the rise and fall of the water level of the ocean.

tide pool (TYD POOL) a temporary pool of ocean water that gets trapped between rocks when the tide goes out.

water pressure (WAWT•er PRESH•er) the downward push of water.

wave (WAYV) the up-and-down movement of surface water.